MATHEMATICAL MODELING OF WING SECTIONS

Boris Dolomanov

Copyright © 2012 by Boris Dolomanov.

ISBN:	Softcover	978-1-4771-0628-0
	Ebook	978-1-4771-0629-7

All rights reserved. No part of this book may be reproduced or transmitted in any form or by any means, electronic or mechanical, including photocopying, recording, or by any information storage and retrieval system, without permission in writing from the copyright owner.

This book was printed in the United States of America.

To order additional copies of this book, contact:
Xlibris Corporation
1-888-795-4274
www.Xlibris.com
Orders@Xlibris.com

About the Brochure

Dear reader!

This brochure is all about the Method of mathematical modeling of wing sections. It is meant for professionals who work on design wings and also blades of propellers or turbines, rudders or any objects that have airfoil in section. The author devoted many years to develop this Method and now brings it to your attention.

When creating this Method, the author tried to make the description simple, avoid complex theoretical structures, strived to convey the main ideas briefly and understandable to anyone who has technical college or university background.

If the reader would like to send their comments on discussed points or report typos, inaccuracies and so on, the author will appreciate.

Please send all comments, Email: dolomanova@msn.com

Boris Dolomanov

Contents

Index of Notation	4
Introduction	5
1. Variation principles of D curves	6
2. Properties of D_n curves	8
3. Mathematical modeling of lines by D_n curves	10
4. Mathematical modeling of lines with compound $D_p(l_j) \oplus D_p(l_k)$ Curves	13
5. Task. Mathematical modeling of wing section for which parameters r, x_M, y_M, x_m, y_m are given	
5.1. Definition of Task	17
5.2. Solution of Task	19
5.3. Equation $H(x_p, s_1, s_2, \beta_1, \beta_2) = 0$	24
Conclusion	25
Bibliography	26
Appendix	27

Index of Notation

$D_n(l_k)$ - n power curve modeling line l_k;

$\gamma(x)$ - the slope angle of tangent to D_n - curve in the point with abscissa x;

$y(x)$ - ordinates function of D_n - curve;

$u(x)$ - function;

$k(x)$, $g(x)$ - functions of curvatures and change curvatures;

$D_p(l_j) \oplus D_q(l_k)$ - compound curve of two D_p and D_q curves modeling lines l_j and l_k;

$S(m)$ - the merge point of two curves with join order equal m;

$x0y$ - system of coordinates;

Γ_1, Γ_2 - the upper and lower surfaces;

$M(x_M, y_M)$ - Maximal point of Γ_1 and its coordinates;

$m(x_m, y_m)$ - minimal point of Γ_2 and its coordinates;

r - radius of curvature at leading edge, point $(0,0)$;

L - trailing edge;

β_1, β_2 - slope angles of tangents to Γ_1 and Γ_2 at point $x = L$;

$P(x_p, y_p)$ - point of lower surface and its coordinates;

$Y(x)$ - main function of ordinates;

$U(x)$ - main function;

$K(x)$ - main function of curvatures;

C_s - inserted in wing section circle with touch points S_1 and S_2;

$(s_1, ys_1), (s_2, ys_2)$ - coordinates of points S_1 and S_2;

$\rho, (\xi_{0s}, \eta_{0s})$ - radius of circle C_s and its coordinates of center;

$\chi(x, s)$ - Heaviside function.

Introduction

Problems of wing's designing which capable to provide the required lift and minimal drag forces always are in focus of aerodynamics. The efforts of scientists are directing to solve two theoretical problems: creation of geometry of wing and definition its dynamics. While the methods of vortex theory allow to solve the second problem, the first is in constant improvement. The main aim of this task is to create a method that will allow to effectively control the wing surface. One of such methods is the Method of mathematical modeling.

1. Variation principles of D curves.

Let's present a curve in the coordinate system $x0y$ that connects point $A(x_1, y_1)$ and $B(x_2, y_2)$. This curve with the axis $0x$ and straight lines $x = x_1$ and $x = x_2$ outlines figure $x_1 ABx_2$ for which:

$$\omega = \int_{x_1}^{x_2} y \, dx, \quad I = \int_{x_1}^{x_2} (x - x_1) y \, dx, \quad J = \int_{x_1}^{x_2} (x - x_1)^2 y \, dx,$$

ω - area of the figure;
I - static moment of area in relation to straight line $x = x_1$;
J - inertia moment of area in relation to straight line $x = x_1$ (see Figure 1);
Let's set $\Phi_3(x_1, y_1, x_2, y_2, \omega, I, J)$, where values included in Φ_3 are known.

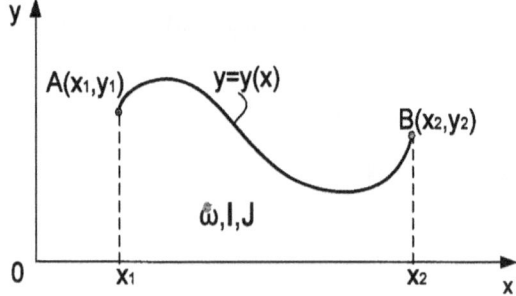

Figure 1. Graph of function connecting points A and B with given $x_1, y_1, x_2, y_2, \omega, I, J$.

Task: Find function $y = y(x)$ which gets the minimum of functional

$$L_D = \int_{x_1}^{x_2} \sqrt{1 + \left[y'(x)\right]^2} \, dx, \tag{1}$$

where L_D - the curve length;
   $y = y(x)$ - continuous differentiable function.
This task belongs to Isoperimetric variation calculation tasks considered in [3]. Required function has form:

$$y(x) = y_1 + \int_{x_1}^{x} F(x)dx = y_1 + \int_{x_1}^{x} \frac{u(x)}{\sqrt{1-u(x)^2}} dx, \quad x \in [x_1, x_2], \quad (2)$$

$$u(x) = a + b(x - x_1) + c(x - x_1)^2 + d(x - x_1)^3, \quad |u(x)| \le 1, \quad (3)$$

where $u(x)$ is the polynomial in power of three, which contains coefficients a, b, c, d. These coefficients are related with Φ_3 by ratios:

$$\int_{x_1}^{x_2} F(x)dx = y_2 - y_1; \quad (4)$$

$$\int_{x_1}^{x_2} (x - x_1) F(x)dx = (x_2 - x_1) y_2 - \omega; \quad (5)$$

$$\int_{x_1}^{x_2} (x - x_1)^2 F(x)dx = (x_2 - x_1)^2 y_2 - 2 \cdot I; \quad (6)$$

$$\int_{x_1}^{x_2} (x - x_1)^3 F(x)dx = (x_2 - x_1)^3 y_2 - 3 \cdot J \quad (7)$$

Thus, the following two tasks can be formulated:
1. Solve the equations (4) - (7) for a given Φ_3 and find coefficients $u(x)$, which determine the ordinates function (2).
2. Set $\Psi_3(x_1, y_1, x_2, a, b, c, d)$, define (2) and then find y_2, ω, I, J from ratios of (4) - (7).

It's not hard to notice that there is a strong correlation between these tasks.
$$(4) - (7)$$
$$y = y(x, \Phi_3) <-> y = y(x, \Psi_3)$$
In the further discussion we are interested in the second task.

Particular cases:
1. $d = 0$. Set $\Psi_2(x_1, y_1, x_2, a, b, c)$ define the function (2), for which $u(x)$-polynomial in power of two. The values of y_2, ω, I are solved from (4) - (6).
2. $c = d = 0$. The function (2) is:

$$y(x) = y_1 + \int_{x_1}^{x} \frac{a + b(x - x_1)}{\sqrt{1 - [a + b(x - x_1)]^2}} dx =$$

$$= y_1 - \frac{1}{b} \left\{ \sqrt{1 - [a + b(x - x_1)]^2} - \sqrt{1 - a^2} \right\},$$

where $\Psi_1(x_1, y_1, x_2, a, b)$ is given. The graph of this function is an arc of circle - shortest curve, which connects points A and B and limiting the area ω. The values of y_2 and ω are found from (4) – (5).

3. $b = c = d = 0$. The function (2) is:

$$y(x) = y_1 + \frac{a}{\sqrt{1 - a^2}} (x - x_1),$$

where $\Psi_0(x_1, y_1, x_2, a)$ is given. The graph of this function is a straight line, shortest curve, which connects points A and B. The value of y_2 is found from (4).

Definition 1: Name the graph of function $y = y(x)$ "D_n- curve", where the index points to the power of $u(x)$, $n \leq 3$.

D_n- curve will be denoted with the index.

2. Properties of D_n curves.

Let's turn to function (2), its derivative

$$y'(x) = F(x) = \frac{u(x)}{\sqrt{1 - u(x)^2}} \qquad (8)$$

1). Not hard to see that

$$u(x) = \sin \gamma(x), \qquad (9)$$

where $\gamma(x)$ – the slope angle of tangent to D_n- curve in the point with abscissa x.

2). Differentiating (8) and with simple transformations we get

$$k(x) = \frac{y''(x)}{\left\{1 + \left[y'(x)\right]^2\right\}^{\frac{3}{2}}} = u'(x), \qquad (10)$$

where $k = k(x)$ – the curvature function.

3). Differentiating (10)

$$g(x) = k'(x) = u''(x),$$

where $g = g(x)$ - the change curvature function.

4). Formulate the rule of the angles and curvatures signs.
   The angle is positive if counter of angle from Ox is against clock-wise; negative if - clock-wise. The curvature is positive, if curve convexity is downwards; it is negative, if the convexity is upwards.

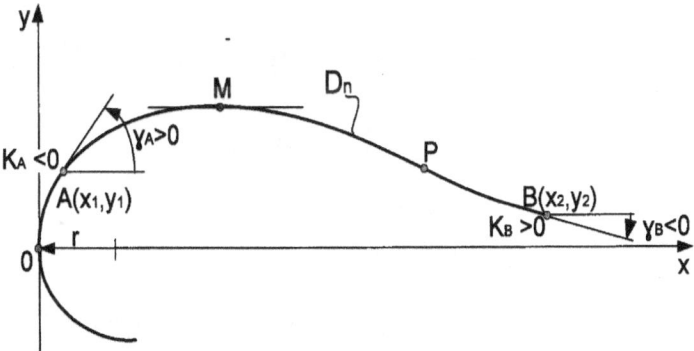

Figure 2. The angles and curvatures signs of D_n- curve.

In Figure 2, the angles and curvatures signs in points A and B are shown. Here are marked the extremal point M for which $\gamma_M = 0$ and the point P, in which curvature $k_p = 0$.

5). If angle $\gamma_0 = \dfrac{\pi}{2}$ then integral in (2) is improper. Define ε - vicinity right of point $x = 0$, then

$$\int_0^x F(x)dx = \int_0^\varepsilon F(x)dx + \int_\varepsilon^x F(x)dx$$

Calculating the first integral we get

$$\int_0^x F(x)dx = \sqrt{2 \cdot r \cdot \varepsilon}\left(1 - \frac{\varepsilon}{2 \cdot r}\right) + \int_\varepsilon^x F(x)dx, \qquad (11)$$

where $\left(\frac{\varepsilon}{2 \cdot r}\right)^2 \ll 1$, $k_0 = -\frac{1}{r} < 0$, r – the curvature radius of D_n- curve in point with abscissa $x = 0$, ε – small value.

6). If angle $\gamma_0 = -\frac{\pi}{2}$, then

$$\int_0^x F(x)dx = -\sqrt{2 \cdot r \cdot \varepsilon}\left(1 - \frac{\varepsilon}{2 \cdot r}\right) + \int_\varepsilon^x F(x)dx, \qquad (12)$$

where $k_0 = \frac{1}{r} > 0$.

7). Let D_n curve of power $n = 3$ intersects axis $0x$ in point $x = 0$ as shown in Figure 2, then for its branch in the upper half-plane $x0y$ the function $u(x)$ is

$$u_+(x) = 1 - \frac{x}{r} + c \cdot x^2 + d \cdot x^3,$$

for the branch in the lower half-plane $x0y$

$$u_-(x) = -1 + \frac{x}{r} + c \cdot x^2 + d \cdot x^3.$$

If $n = 2$ then in these expressions $d = 0$, if $n = 1$ then $c = d = 0$.

8). The length AB of D_n curve is calculated by formula

$$L_{AB} = \int_{x_1}^{x_2} \sqrt{1 + [y'(x)]^2}\, dx = \int_{x_1}^{x_2} \frac{1}{\sqrt{1 - u(x)^2}}\, dx, \qquad (13)$$

3. Mathematical modeling of lines by D_n curves.

Let's define the terms first.
We assume that the object of modeling is line l, and the result of modeling is

D_n curve.

Definition 2: Mathematical model of a line l is ordinates function $y = y(x)$ of $D_n(l)$ curve and the bordering conditions that allow to find coefficients of polynomial $u(x)$, in which case the bordering conditions reflect the line geometry.

Definition 3: Mathematical modeling is the process of building the mathematical model.

Let's illustrate these definitions and theoretical conclusions of the previous paragraphs on a concrete example.

Example 1. Let's depict in Figure 3 a Chinese sign "Ying and Yang" which presents a circle with a line, where line l is like sinusoid.

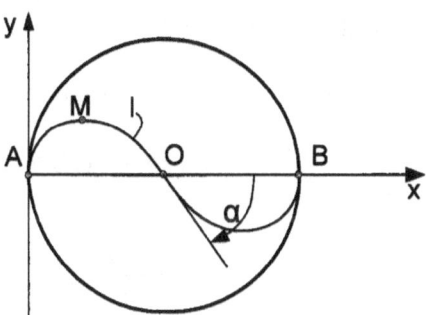

Figure 3. Sign "Ying and Yang".

Offer a few designers to build line l. May say that every one of them will create his own image of the sign with a different position of the extreme point M and the value of angle α.

Let's resolve this task via the Method of modeling line l by $D_n(l)$ - curve.

Let's connect system of coordinates xAy with circle as shown in Figure 3. Because line l is symmetrical to O, let's consider only its left part. The object of modeling is line l.

Assume that the radius of circle is $r = 1$.

The ordinates function of $D_n(l)$ curve looks like:

$$y(x) = y(0) + \sqrt{2 \cdot \varepsilon}\left(1 - \frac{\varepsilon}{2}\right) + \int_{\varepsilon}^{x} \frac{u(x)}{\sqrt{1 - u(x)^2}} dx, \quad x \in [0,1] \quad (14)$$

Boundary conditions of line l in points A and O
$$y(0) = 0, \quad u(0) = 1, \quad k(0) = -1, \quad (15), (16), (17),$$
$$y(1) = 0, \quad u(1) = \sin\alpha, \quad k(1) = 0, \quad (18), (19), (20),$$
where angle α is unknown.

Boundary conditions are for calculating the coefficients of function $u(x)$, where the value of power of $u(x)$ is defined by the number of boundary conditions. Find this value. Condition (15) is considered in (14), condition (18) is necessary to find the angle α. Thus, four conditions (16), (17) and (19), (20) are used to define four coefficients of function $u(x)$, which has power of three. Using (16), (17), write

$$u(x) = 1 - x + c \cdot x^2 + d \cdot x^3, \quad k(x) = -1 + 2 \cdot c \cdot x + 3 \cdot d \cdot x^2$$

The conditions (19), (20) allow to build simple equations

$$c + d = \sin\alpha, \quad 2 \cdot c + 3 \cdot d = 1,$$

then

$$c = c(\alpha) = -1 + 3\sin\alpha, \quad d = d(\alpha) = 1 - 2\sin\alpha$$

The angle α is found from the equation

$$y(1,\alpha) = \sqrt{2 \cdot \varepsilon}\left(1 - \frac{\varepsilon}{2}\right) + \int_{\varepsilon}^{1} \frac{u(x,\alpha)}{\sqrt{1 - u(x,\alpha)^2}} dx = 0,$$

where
$$u(x,\alpha) = 1 - x + c(\alpha)x^2 + d(\alpha)x^3$$

Mathematical modeling is completed by building $D_3(l)$ - curve and the print of table of its points coordinates.

Program "Sign Ying and Yang" implements the solution of this task (see Appendix of brochure). In Fig.1, $D_3(l)$ curve is built, the value of angle α and coordinates of point M are calculated.

The author offers to reader to assess the "correctness" of $D_3(l)$ curve.

Note: All calculations of the tasks completed with the help of Mathcad - genius system made by Mathsoft.

4. Mathematical modeling of lines with compound $D_p(l_j) \oplus D_q(l_k)$ curves.

Let's understand under a "complex" line l, mathematical modeling of which is not realized by only one D_n curve. In this case the complex line should be presented as consisting of several parts.

Consider line $l(AB)$ that consists of lines l_1 and l_2 as shown in Figure 4. These lines modeling with curves $D_p(l_1)$ and $D_q(l_2)$, which merge in point $S(m)$, where m is the order of merging. Their geometrical sum is defined in the form $D_p(l_1) \oplus D_q(l_2)$ and call it "compound" curve. By the order of merging m we understand the conditions which must be satisfied for functions of curves $D_p(l_1)$ and $D_q(l_2)$ in point S. These conditions are presented in Table 1.

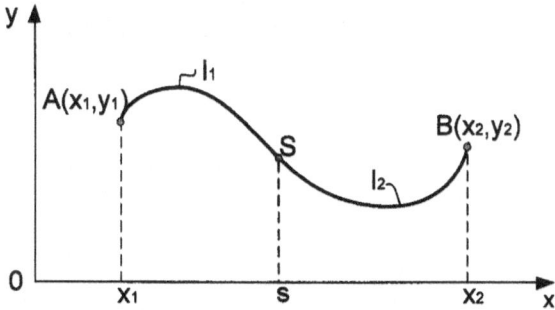

Figure 4. Complex line that consists of two parts.

The mathematical model of complex line contains the ordinates functions of $D_p(l_1)$ and $D_q(l_2)$ curves, their boundary conditions and conditions of merging. These conditions must allow to find coefficients of functions $u_1(x), u_2(x)$ and the abscissa s at point S if it is not given.

Notes:
1). The order of merging is set from the requirements to smoothness of compound curve in point S.
2). The abscissa of point S is better not to set but let the mathematical Method find its value. In this case, "squeezing" of curve $D_p(l_1)$ and $D_q(l_2)$ must be

excluded.

3). If each $D_p(l_1)$ and $D_q(l_2)$ curve according to definition bring the minimum of functional (1), then the $D_p(l_1) \oplus D_q(l_2)$ curve in general case does not have this property.

Table 1

The order of merging m	The conditions of merging in point S
0	$y_1(s) = y_2(s)$
1	$y_1(s) = y_2(s)$, $u_1(s) = u_2(s)$
2	$y_1(s) = y_2(s)$, $u_1(s) = u_2(s)$, $k_1(s) = k_2(s)$
3	$y_1(s) = y_2(s)$, $u_1(s) = u_2(s)$, $k_1(s) = k_2(s)$, $g_1(s) = g_2(s)$

Let's consider an example of how the modeling of complex line is done.

Example 2. Find the axis section of rotation body having the cylinder insertion.
For body are given: length $OL = 1$, diameter d and angle β in point L

Let's name d and β as "parameters of object" or simply "parameters". At the first glance, it may seem that two parameters may not be enough, but as it will be shown later, the first impression is wrong.
If this task is given to a designer, then he will easily build a section shown in Figure 5. However, when asked: "How points S_1 and S_2 are located?" will probably not be able to give a proved answer. Based on these considerations, the

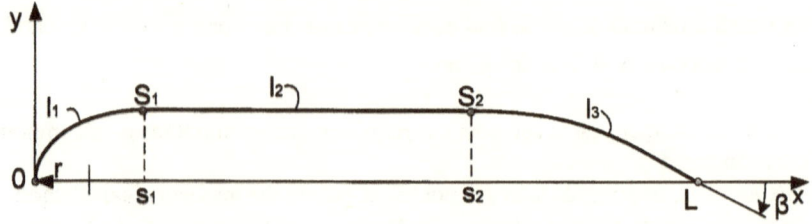

Figure 5. Axis section of rotation body.

designer's blueprint should be simply viewed as a sketch.
It's not hard to see that the line of section is complex. Let's divide this line into parts l_1, l_2 and l_3, which we modeling by compound curve
$$D_2(l_1) \oplus D_0(l_2) \oplus D_3(l_3)$$
The merge points are $S_1(2)$ and $S_2(2)$. Writing the boundary conditions and the merge conditions in points $0, S_1, S_2$ and L:

$x = 0 \qquad y_1(0) = 0, \quad u_1(0) = 1, \quad k_1(0) = -\dfrac{1}{r},$ \hfill (21), (22), (23),

$x = s_1 \quad y_1(s_1) = y_2 = \dfrac{d}{2}, \quad u_1(s_1) = 0, \quad k_1(s_1) = 0,$ \hfill (24), (25), (26), (27),

$x = s_2 \qquad y_2 = y_3(s_2) = \dfrac{d}{2}, \quad u_3(s_2) = 0,$ \hfill (28), (29), (30),

$\qquad\qquad\qquad k_3(s_2) = 0,$ \hfill (31),

$x = 1 \quad y_3(1) = 0, \quad u_3(1) = \sin \beta, \quad k_3(1) = 0$ \hfill (32), (33), (34),

where r - unknown radius in point 0;
  s_1, s_2 - abscissas of points S_1 и S_2 are unknown.

The functions of curves have forms:

$D_2(l_1): \quad y_1(x,r) = \sqrt{2 \cdot r \cdot \varepsilon}\left(1 - \dfrac{\varepsilon}{2 \cdot r}\right) + \displaystyle\int_\varepsilon^x \dfrac{u_1(x,r)}{\sqrt{1 - u_1(x,r)^2}} dx \quad x \in [0, s_1],$

$u_1(x,r) = 1 - \dfrac{x}{r} + c_1 \cdot x^2, \quad k_1(x,r) = -\dfrac{1}{r} + 2 \cdot c_1 \cdot x,$

where conditions of (21) – (23) are accounted. Conditions of (26) and (27) allow to get the equations
$$1 - \dfrac{s_1}{r} + c_1 \cdot s_1^2 = 0, \qquad -\dfrac{1}{r} + 2 \cdot c_1 \cdot s_1 = 0$$
Solving these equations we get
$$s_1 = 2 \cdot r \text{ and } c_1 = c_1(r) = \dfrac{1}{4 \cdot r^2}, \text{ then } u_1(x,r) = \left(1 - \dfrac{x}{2 \cdot r}\right)^2.$$
We define r using (24). This condition produces the equation

$$y_1(2 \cdot r, r) - \frac{d}{2} = 0$$

Horizontal part of section modeling by curve

$$D_0(l_2): \quad y_2 - \frac{d}{2} = 0, \quad x \in [s_1, s_2]$$

$$D_3(l_3): \quad y_3(x, s_2) = \frac{d}{2} + \int_{s_2}^{x} \frac{u_3(x, s_2)}{\sqrt{1 - u_3(x, s_2)^2}} dx, \quad x \in [s_2, 1],$$

$$u_3(x, s_2) = c_3(x - s_2)^2 + d_3(x - s_2)^3,$$

$$k_3(x, s_2) = 2 \cdot c_3(x - s_2) + 3 \cdot d_3(x - s_2)^2,$$

where the conditions (30) - (31) are taken into account. Using (33) and (34) we get the equations.

$$c_3(1 - s_2)^2 + d_3(1 - s_2)^3 = \sin \beta, \quad 2 \cdot c_3(1 - s_2) + 3 \cdot d_3(1 - s_2)^2 = 0$$

Formulas for coefficients of functions $u_3(x, s_2)$ are

$$c_3 = c_3(s_2) = 3 \cdot f(s_2), \quad d_3 = d_3(s_2) = -\frac{2}{1 - s_2} f(s_2), \quad f(s_2) = \frac{\sin \beta}{(1 - s_2)^2}$$

Condition (32) allows to get the equation for definition of abscissa of s_2

$$y_3(1, s_2) = 0$$

Writing the functions of compound curve.

$$Y(x) = \begin{vmatrix} y_1(x, r), x \in [0, s_1); \\ \dfrac{d}{2}, x \in [s_1, s_2); \\ y_3(x, s_2), x \in [s_2, 1]. \end{vmatrix} \qquad U(x) = \begin{vmatrix} u_1(x, r), x \in [0, s_1); \\ 0, x \in [s_1, s_2); \\ u_3(x, s_2), x \in [s_2, 1]. \end{vmatrix}$$

$$K(x) = \frac{d}{dx} U(x)$$

Functions $Y(x), U(x), K(x)$ in further called "main functions".

We introduce the program "Rotation body" which is used to calculate the section with diameter $d = 0.15$ and angle $\beta = -\dfrac{\pi}{8}$.

Note: If we assume that l_1 is an arc of circle (nose of rotation body – half- sphere), then in point S_1 the curvature has a break. When the body moves this is cause of the jump of centripetal acceleration of the flow particles. This unwanted phenomenon reproduces premature turbulence in the boundary layer and increase of resistance.

5. Task. Mathematical modeling of wing section for which parameters r, x_M, y_M, x_m, y_m are given.

5.1. Definition of Task.

Let's try to consider all stages from definition of the task to the last formula of its solution.
The definition of the task includes: description of modeling schema and boundary conditions.
Let's draw a sketch of wing section and associate it with a system of coordinates $x0y$ so that the $0x$ axis goes through a point L and maximal remote point 0 in the nose of section. The axis of $0y$ is pointed up and is perpendicular to axis $0x$. Assume that the length of section – is the distance $0L$ equal to one. The axis $0x$ divides the section into the upper Γ_1 and the lower Γ_2 surfaces. Let's denote on Γ_1 and Γ_2 extreme points $M(x_M, y_M)$ and $m(x_m, y_m)$. Parameters of the wing section are the coordinates of M and m and also r – radius of curvature in point 0.
Let's divide Γ_1 with point S_1 into two lines l_1 and l_3. Assume that the Γ_2 consists of three lines l_0, l_2, l_4, which are divided by points P and S_2. The positions of points P, S_1 and S_2 on the section are unknown. Modeling Γ_1 with compound curve $D_3(l_1) \oplus D_2(l_3), \Gamma_2 - D_1(l_0) \oplus D_3(l_2) \oplus D_2(l_4)$, where curve $D_1(l_0)$ - is the arc of circle of radius r. Let's for each point set the merge order $O(2), P(2), S_1(3), S_2(3), L(0)$.
All mentioned settings are shown on Schema of Modeling.
The task is solved if the functions of all D_n - curves is defined. For this we need

Schema of Modeling.

1. Wing section sketch.
   Given parameters: r, x_M, y_M, x_m, y_m

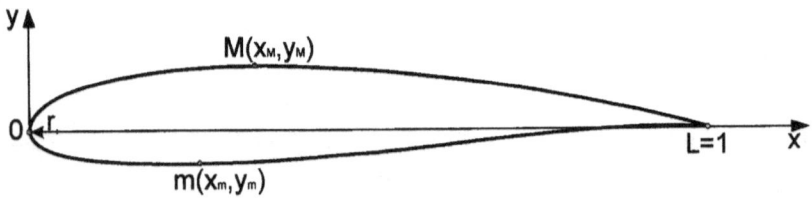

2. Lines of section and merge points.

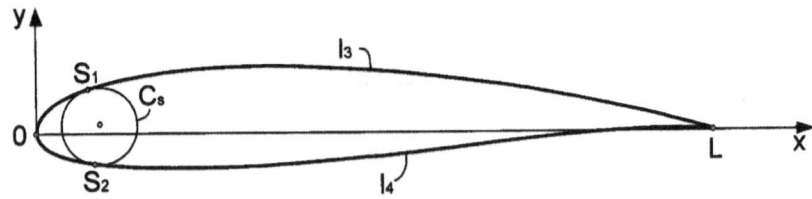

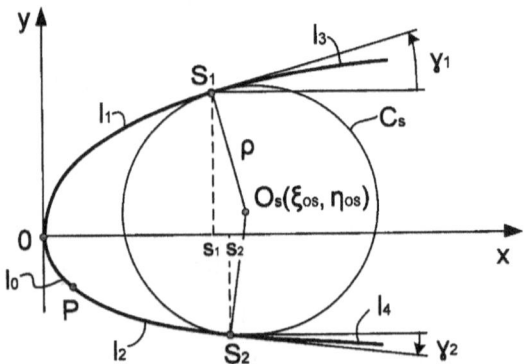

3. Compound curves:
   upper surface - $D_3(l_1) \oplus D_2(l_3)$;
   lower surface - $D_1(l_0) \oplus D_3(l_2) \oplus D_2(l_4)$
4. The merge order in points:
   $O(2), P(2), S_1(3), S_2(3), L(0)$

to find all unknowns of these functions.
Boundary conditions.
Upper surface

$$
\begin{array}{lll}
x = 0 & y_1(0) = 0, \quad u_1(0) = 1, \quad k_1(0) = -\dfrac{1}{r}, & \text{(T.1), (T.2), (T.3),} \\
x = s_1 & y_1(s_1) = y_3(s_1), \quad u_1(s_1) = u_3(s_1), & \text{(T.4), (T.5),} \\
 & k_1(s_1) = k_3(s_1), \quad g_1(s_1) = g_3(s_1), & \text{(T.6), (T.7),} \\
x = x_M & y_3(x_M) = y_M, \quad u_3(x_M) = 0, & \text{(T.8), (T.9),} \\
x = 1 & y_3(1) = 0, & \text{(T.10),}
\end{array}
$$

where s_1 - abscissa of point S_1.
Lower surface:

$$
\begin{array}{lll}
x = 0 & y_0(0) = 0, \quad u_0(0) = -1, \quad k_0(0) = \dfrac{1}{r}, & \text{(T.11), (T.12), (T.13),} \\
x = x_p & y_0(x_p) = y_2(x_p), \quad u_0(x_p) = u_2(x_p), & \text{(T.14), (T.15),} \\
 & \dfrac{1}{r} = k_2(x_p), & \text{(T.16),} \\
x = s_2 & y_2(s_2) = y_4(s_2), \quad u_2(s_2) = u_4(s_2), & \text{(T.17), (T.18),} \\
 & k_2(s_2) = k_4(s_2), \quad g_2(s_2) = g_4(s_2), & \text{(T.19), (T.20),} \\
x = x_m & y_4(x_m) = y_m, \quad u_4(x_m) = 0, & \text{(T.21), (T.22),} \\
x = 1 & y_4(1) = 0, & \text{(T.23),}
\end{array}
$$

where x_p, s_2 - abscissas of points P, S_2.

5.2. Solution of Task.

The functions of the upper surface curves are forms:

$$D_3(l_1): \quad y_1(x) = \sqrt{2 \cdot r \cdot \varepsilon}\left(1 - \dfrac{\varepsilon}{2 \cdot r}\right) + \int_\varepsilon^x \dfrac{u_1(x)}{\sqrt{1 - u_1(x)^2}} dx, \quad x \in [0, s_1],$$

$$u_1(x) = 1 - \frac{x}{r} + c_1 x^2 + d_1 x^3, \qquad k_1(x) = -\frac{1}{r} + 2 \cdot c_1 x + 3 \cdot d_1 x^2,$$

$$g_1(x) = 2 \cdot c_1 + 6 \cdot d_1 x,$$

where the conditions (T.1) – (T.3) are taken into consideration.

$$D_2(l_3): \quad y_3(x) = y_1(s_1) + \int_{s_1}^{x} \frac{u_3(x)}{\sqrt{1 - u_3(x)^2}} dx, \qquad x \in [s_1, 1],$$

$$u_3(x) = a_3 + b_3(x - s_1) + c_3(x - s_1)^2, \quad k_3(x) = b_3 + 2 \cdot c_3(x - s_1), \quad g_3 = 2 \cdot c_3,$$

where for ordinates function the condition (T.4) is included. Let's convert function $u_3(x)$ using (T.5) - (T.7).

$$u_3(x) = 1 - \frac{s_1}{r} + c_1 s_1^2 + d_1 s_1^3 +$$

$$+ \left(-\frac{1}{r} + 2 \cdot c_1 s_1 + 3 \cdot d_1 s_1^2\right) \cdot (x - s_1) + (c_1 + 3 \cdot d_1 s_1) \cdot (x - s_1)^2 =$$

$$= 1 - \frac{x}{r} + c_1 x^2 + d_1 \left[x^3 - (x - s_1)^3\right] = u_1(x) - d_1(x - s_1)^3$$

Unknowns are coefficients c_1, d_1 and abscissa s_1. Let's enter additional condition.

$$u_3(1) = \sin \beta_1, \tag{T.24}$$

where β_1 – is the slope angle of the tangent to Γ_1 in point L, which is unknown. Conditions (T.9) and (T.24) allow writing the equations.

$$\begin{cases} 1 - \frac{x_M}{r} + c_1 x_M^2 + d_1 \left[x_M^3 - (x_M - s_1)^3\right] = 0, \\ 1 - \frac{1}{r} + c_1 + d_1 \left[1 - (1 - s_1)^3\right] = \sin \beta_1 \end{cases}$$

Solving this system of equations we find

$$d_1 = d_1(s_1, \beta_1) = -\frac{1 - \frac{x_M}{r} - \left(1 - \frac{1}{r} - \sin \beta_1\right) \cdot x_M^2}{\mu_1(s_1) - \left[1 - (1 - s_1)^3\right] \cdot x_M^2},$$

$$c_1 = c_1(s_1, \beta_1) = -\frac{1}{x_M^2}\left[1 - \frac{x_M}{r} + d_1(s_1, \beta_1) \cdot \mu_1(s_1)\right],$$

where $\mu_1(s_1) = x_M^3 - (x_M - s_1)^3$

To find s_1 and β_1 we need two equations.

The functions of the lower surface curves.

$D_1(l_0)$: $y_0(x) = -\sqrt{r^2 - (r-x)^2}$, $u_0(x) = -1 + \dfrac{x}{r}$, $k_0 = \dfrac{1}{r}$, $x \in [0, x_p]$

The arc of circle $D_1(l_0)$ has the center in point $(r,0)$. For this curve we complete conditions (T.11) - (T.13).

$D_3(l_2)$: $y_2(x) = y_0(x_p) + \displaystyle\int_{x_p}^{x} \dfrac{u_2(x)}{\sqrt{1 - u_2(x)^2}} dx$, $x \in [x_p, s_2]$,

$$u_2(x) = -1 + \dfrac{x}{r} + c_2(x - x_p)^2 + d_2(x - x_p)^3,$$

$k_2(x) = \dfrac{1}{r} + 2 \cdot c_2(x - x_p) + 3 \cdot d_2(x - x_p)^2$, $g_2(x) = 2 \cdot c_2 + 6 \cdot d_2(x - x_p)$

Functions of curve $D_3(l_2)$ written with regard of conditions (T.14) and (T.16).

$D_2(l_4)$: $y_4(x) = y_2(s_2) + \displaystyle\int_{s_2}^{x} \dfrac{u_4(x)}{\sqrt{1 - u_4(x)^2}} dx$, $x \in [s_2, 1]$,

$u_4(x) = a_4 + b_4(x - s_2) + c_4(x - s_2)^2$, $k_4(x) = b_4 + 2 \cdot c_4(x - s_2)$, $g_4 = 2 \cdot c_4$.
For functions $y_4(x)$ we consider (T.17). Let's complete $u_4(x)$ using conditions (T.18) - (T.20).

$$u_4(x) = -1 + \dfrac{s_2}{r} + c_2(s_2 - x_p)^2 + d_2(s_2 - x_p)^3 +$$

$$+ \left[\dfrac{1}{r} + 2 \cdot c_2(s_2 - x_p) + 3 \cdot d_2(s_2 - x_p)^2 \right] \cdot (x - s_2) +$$

$$+ \left[c_2 + 3 \cdot d_2(s_2 - x_p) \right] \cdot (x - s_2)^2 =$$

$$= -1 + \dfrac{x}{r} + c_2(x - x_p)^2 + d_2 \left[(x - x_p)^3 - (x - s_2)^3 \right] = u_2(x) - d_2(x - s_2)^3$$

Coefficients c_2, d_2 and abscissas x_p, s_2 are unknown. Enter additional condition

$$u_4(1) = \sin \beta_2, \qquad (T.25)$$

in which β_2 is the slope angle of the tangent to Γ_2 in point L is unknown.
Conditions (T.22) and (T.25) allow us to write the equations.

$$\begin{cases} -1+\dfrac{x_m}{r}+c_2(x_m-x_p)^2+d_2\left[(x_m-x_p)^3-(x_m-s_2)^3\right]=0, \\ -1+\dfrac{1}{r}+c_2(1-x_p)^2+d_2\left[(1-x_p)^3-(1-s_2)^3\right]=\sin\beta_2 \end{cases}$$

Solving this system of equations we find the formulas.

$$d_2 = d_2(x_p, s_2, \beta_2) = \dfrac{1-\dfrac{x_m}{r}-\left(1-\dfrac{1}{r}+\sin\beta_2\right)\cdot\lambda_2(x_p)}{\mu_2(x_p, s_2)-\left[(1-x_p)^3-(1-s_2)^3\right]\cdot\lambda_2(x_p)},$$

$$c_2 = c_2(x_p, s_2, \beta_2) = \dfrac{1}{(x_m-x_p)^2}\left[1-\dfrac{x_m}{r}-d_2(x_p, s_2, \beta_2)\cdot\mu_2(x_p, s_2)\right],$$

where $\mu_2(x_p, s_2) = (x_m-x_p)^3-(x_m-s_2)^3$, $\lambda_2(x_p) = \left(\dfrac{x_m-x_p}{1-x_p}\right)^2$

To find x_p, s_2 and β_2 we need three equations.

The equations system of task.

1) Conditions (T.8), (T.10) and (T.21), (T.23) allow to write four equations.

$$\begin{cases} y_3(x_M, s_1, \beta_1)-y_M = 0, \\ y_3(1, s_1, \beta_1) = 0, \\ y_4(x_m, x_p, s_2, \beta_2)-y_m = 0, \\ y_4(1, x_p, s_2, \beta_2) = 0 \end{cases} \quad (T.26)$$

We need the fifth equation.

Let's enter circle C_s into the wing section as shown in Schema of Modeling. The circle radius is denoted by ρ and the coordinates of the center (ξ_{0s}, η_{0s}).

Hypothesis: S_1, S_2 are the touching points of Γ_1, Γ_2 and circle C_s.

The outcome of the hypothesis is the equation:

$$H(x_p,s_1,s_2,\beta_1,\beta_2) = s_2 - s_1 + \frac{u_2(s_2,x_p,s_2,\beta_2) + u_1(s_1,s_1,\beta_1)}{\sqrt{1-u_2(s_2,x_p,s_2,\beta_2)^2} + \sqrt{1-u_1(s_1,s_1,\beta_1)^2}} \times$$
$$\times \left[y_2(s_2,x_p,s_2,\beta_2) - y_1(s_1,s_1,\beta_1) \right] = 0, \qquad (\text{T.27})$$

which we will get in 5.3.
Solving joint equations (T.26) and (T.27) we find the unknowns: $x_p, s_1, s_2, \beta_1, \beta_2$.

The main functions of wing section.

The upper surface: $\quad Y_1(x) = \begin{vmatrix} y_1(x,s_1,\beta_1), x \in [0,s_1); \\ y_3(x,s_1,\beta_1), x \in [s_1,1]. \end{vmatrix}$

$$U_1(x) = u_1(x,s_1,\beta_1) - d_1(s_1,\beta_1)(x-s_1)^3 \chi_1(x,s_1),$$
$$K_1(x) = k_1(x,s_1,\beta_1) - 3 \cdot d_1(s_1,\beta_1)(x-s_1)^2 \chi_1(x,s_1).$$

The lower surface: $\quad Y_2(x) = \begin{vmatrix} y_0(x), x \in [0,x_p); \\ y_2(x,x_p,s_2,\beta_2), x \in [x_p,s_2); \\ y_4(x,x_p,s_2,\beta_2), x \in [s_2,1] \end{vmatrix}$

$$U_2(x) = \begin{vmatrix} u_0(x), x \in [0,x_p); \\ u_2(x,x_p,s_2,\beta_2) - d_2(x_p,s_2,\beta_2)(x-s_2)^3 \chi_2(x,s_2), x \in [x_p,1]. \end{vmatrix}$$

$$K_2(x) = \begin{vmatrix} k_0, x \in [0,x_p); \\ k_2(x,x_p,s_2,\beta_2) - 3 \cdot d_2(x_p,s_2,\beta_2)(x-s_2)^2 \chi_2(x,s_2), x \in [x_p,1], \end{vmatrix}$$

where $\chi_1(x,s_1)$ and $\chi_2(x,s_2)$ - are the Heaviside functions.

5.3. Equation $H = H(x_p,s_1,s_2,\beta_1,\beta_2)$.

It's easy to see that equalities are fair

$$s_1 = \xi_{0s} - \rho \cdot \sin \gamma_1 = \xi_{0s} - \rho \cdot u_1(s_1), \qquad (T.28)$$
$$s_2 = \xi_{0s} + \rho \cdot \sin \gamma_2 = \xi_{0s} + \rho \cdot u_2(s_2), \qquad (T.29)$$

where γ_1, γ_2 - the slope angles of tangents to Γ_1 and Γ_2 in points S_1, S_2. Subtracting from (T.29) equality (T.28) we receive

$$\rho(s_1, s_2) = \frac{s_2 - s_1}{u_2(s_2) + u_1(s_1)} \qquad (T.30)$$

Let's write also

$$y_1(s_1) = \eta_{0s} + \rho \cdot \cos \gamma_1 = \eta_{0s} + \rho\sqrt{1 - u_1(s_1)^2}, \qquad (T.31)$$
$$y_2(s_2) = \eta_{0s} - \rho \cdot \cos \gamma_2 = \eta_{0s} - \rho\sqrt{1 - u_2(s_2)^2} \qquad (T.32)$$

If we subtract from (T.32) equality (T.31) we get

$$\rho(s_1, s_2) = -\frac{y_2(s_2) - y_1(s_1)}{\sqrt{1 - u_2(s_2)^2} + \sqrt{1 - u_1(s_1)^2}} \qquad (T.33)$$

The right parts (T.30) and (T.33) are equal, that allows to receive

$$H(s_1, s_2) = s_2 - s_1 + \frac{u_2(s_2) + u_1(s_1)}{\sqrt{1 - u_2(s_2)^2} + \sqrt{1 - u_1(s_1)^2}}[y_2(s_2) - y_1(s_1)] = 0 \quad (T.34)$$

This equation connects abscissas of merging points S_1 and S_2.
The functions of curves $D_3(l_1)$ and $D_3(l_2)$ have forms:

$$y_1(s_1) = y_1(s_1, s_1, \beta_1), \quad u_1(s_1) = u_1(s_1, s_1, \beta_1),$$
$$y_2(s_2) = y_2(s_2, x_p, s_2, \beta_2), \quad u_2(s_2) = u_2(s_2, x_p, s_2, \beta_2)$$

Thus equation (T.34) needs to record as $H(x_p, s_1, s_2, \beta_1, \beta_2) = 0$.

Conclusion

The brochure contains basic ideas of the mathematical modeling Method of wing sections.
Let's list the stages of realization of this Method.

1). *The sketch* reflects all features of the form of wing section and contains parameters, opening their geometrical sense.
2). *The scheme of modeling* contains lines of wing section – separate sites of the upper and lower surface. These lines are appointed so that their modeling could be executed by D_n curves of the power $n \leq 3$.

    Compound curves are geometrical sum D_n curves. The upper and lower surface are modeled by compound curves. Points of merging limit D_n curves. The order of each merge point is appointed, $m \leq 3$.
3). *Boundary conditions* are the mathematical requirements to functions of D_n curves.
    Boundary conditions: conditions of wing section parameters and conditions in points of merging. Number unknown functions equally number of boundary conditions.
4). *The solution of problem* is a finding of all unknown by the decision of the equations, received from boundary conditions.
5). *The engineering specifications*: Manufacturing the drawings of wing section and its fragments and printing the calculated tables.

Notes:

> The author is not familiar with other methods of generating the wing sections for which only the length, the coordinates of two points and the nose radius are given.
> The author will appreciate if the reader points to existence of other more efficient methods of decision this task.

Dear reader, I wish you good luck!

Bibliography

1. Бронштейн И.Н., Семендяев К. А. Справочник по математике, Москва, " Наука", 1968.
2. Гурский Д. А. Вычисления в MathCAD, Минск, ООО "Новое знание", 2003.
3. Смирнов В. И. Курс высшей математики, Москва, "Наука", 1974.
4. Ira H. Abbott, Albert E. von Doenhoff. Theory of wing sections. Dover publications Inc, New York.
5. Foux L.D., Pratt M.J. Computational geometry for design and manufacture, John Wiley & Sons, New York.

Appendix

6. Program "Sign Ying and Yang"...28
7. Program "Rotation body"...29
8. Program "Wing section"...31
9. Calculations of wing sections..36

6. PROGRAM "Sign Ying and Yang"

1. Parameter: $\quad r := 1$

2. D3(L) $\quad c(\alpha) := -1 + 3 \cdot \sin(\alpha) \quad d(\alpha) := 1 - 2 \cdot \sin(\alpha)$

$u(x,\alpha) := 1 - x + c(\alpha) \cdot x^2 + d(\alpha) \cdot x^3 \quad k(x,\alpha) := -1 + 2 \cdot c(\alpha) \cdot x + 3 \cdot d(\alpha) \cdot x^2$

Function of Ordinates $\quad \varepsilon := 10^{-4} \quad y(x,\alpha) := \sqrt{2 \cdot \varepsilon} \cdot \left(1 - \frac{\varepsilon}{2}\right) + \int_{\varepsilon}^{x} \frac{u(x,\alpha)}{\sqrt{1 - u(x,\alpha)^2}} dx$

3. Angel α $\quad \alpha := -\frac{\pi}{3} \quad \alpha := \text{root}(y(1,\alpha),\alpha) \quad \alpha = -1.2803$

4. Coordinates of Point M $\quad xM := 0.5 \quad xM := \text{root}(u(xM,\alpha),xM) \quad yM := y(xM,\alpha)$

$\qquad xM = 0.4613 \quad yM = 0.6396$

5. Circle C $\quad \xi(\theta) := 1 + \cos(\theta) \quad \eta(\theta) := \sin(\theta) \quad \theta := 0, \frac{\pi}{50} .. 2 \cdot \pi$

6. Sign Ying and Yang $\quad Y(x) := \begin{vmatrix} 0 & \text{if } y(x,\alpha) < 10^{-5} \\ y(x,\alpha) & \text{otherwise} \end{vmatrix} \quad x := 0, 0.01 .. 1$

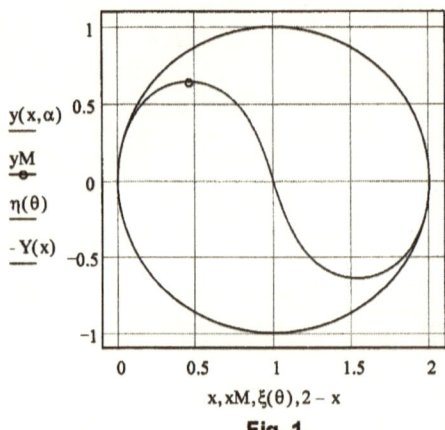

Fig. 1

$x := 0, 0.1 .. 1$

x	Y(x)
0	0
0.1	0.4108
0.2	0.5355
0.3	0.6027
0.4	0.6344
0.5	0.6376
0.6	0.6131
0.7	0.557
0.8	0.4582
0.9	0.2883
1	0

MATHEMATICAL MODELING OF WING SECTIONS

7. PROGRAM "Rotation body"

1. Parameters: $\quad d := 0.15 \quad \beta := -\dfrac{\pi}{8}$

2. D2(L1)

$$u1(x,r) := \left(1 - \dfrac{x}{2 \cdot r}\right)^2 \quad \varepsilon := 10^{-4} \quad y1(x,r) := \sqrt{2 \cdot r \cdot \varepsilon} \cdot \left(1 - \dfrac{\varepsilon}{2 \cdot r}\right) + \int_{\varepsilon}^{x} \dfrac{u1(x,r)}{\sqrt{1 - u1(x,r)^2}} \, dx$$

$r := 0.1 \quad r := \text{root}\left(y1(2 \cdot r, r) - \dfrac{d}{2}, r\right) \quad r := \text{root}\left(y1(2 \cdot r, r) - \dfrac{d}{2}, r\right) \quad r = 0.06254 \quad s1 := 2 \cdot r$

3. D0(L2) $\quad\quad\quad\quad\quad\quad\quad\quad y2 := \dfrac{d}{2}$

4. D3(L3) $\quad\quad f(s2) := \dfrac{\sin(\beta)}{(1-s2)^2} \quad c3(s2) := 3 \cdot f(s2) \quad d3(s2) := -\dfrac{2}{1-s2} \cdot f(s2)$

$$u3(x,s2) := c3(s2) \cdot (x - s2)^2 + d3(s2) \cdot (x - s2)^3$$

$$y3(x,s2) := \dfrac{d}{2} + \int_{s2}^{x} \dfrac{u3(x,s2)}{\sqrt{1 - u3(x,s2)^2}} \, dx$$

$s2 := 0.6 \quad s2 := \text{root}(y3(1, s2), s2) \quad s2 = 0.6264$

5. Main Function $\quad x := 0, 0.001 .. 1 \quad y(x) := \begin{vmatrix} y1(x,r) & \text{if } 0 \le x < s1 \\ y2 & \text{if } s1 \le x \le s2 \\ y3(x,s2) & \text{if } s2 < x < 1 \\ 0 & \text{if } x = 1 \end{vmatrix}$

$$Y(x) := \begin{vmatrix} 0 & \text{if } y(x) < 10^{-5} \\ y(x) & \text{otherwise} \end{vmatrix}$$

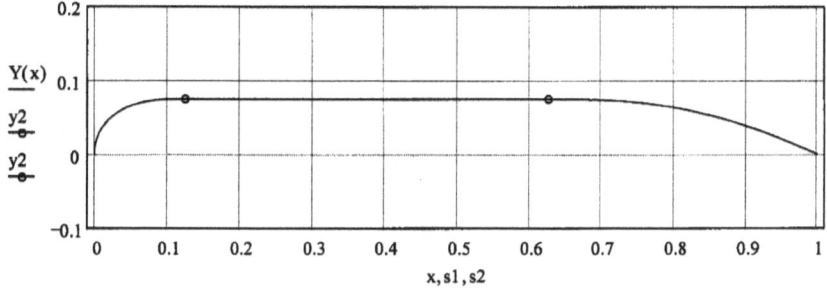

6. Coordinates of Points

$x := 0, \frac{s1}{20} .. s1$

x	Y(x)
0	0
0.0063	0.0274
0.0125	0.0379
0.0188	0.0454
0.025	0.0513
0.0313	0.056
0.0375	0.0598
0.0438	0.063
0.05	0.0657
0.0563	0.0679
0.0625	0.0697
0.0688	0.0711
0.0751	0.0723
0.0813	0.0732
0.0876	0.0738
0.0938	0.0743
0.1001	0.0746
0.1063	0.0748
0.1126	0.0749
0.1188	0.075
0.1251	0.075

$x := s2, s2 + \frac{1 - s2}{20} .. 1$

x	Y(x)
0.6264	0.075
0.6451	0.075
0.6638	0.0749
0.6825	0.0746
0.7012	0.074
0.7198	0.073
0.7385	0.0717
0.7572	0.0699
0.7759	0.0677
0.7945	0.0648
0.8132	0.0615
0.8319	0.0576
0.8506	0.0531
0.8693	0.048
0.8879	0.0424
0.9066	0.0363
0.9253	0.0297
0.944	0.0226
0.9626	0.0153
0.9813	0.0077
1	0

8. Program "Wing section".

Program contains 10 parts:
1). Enter parameters: r, x_M, y_M, x_m, y_m.
2). Writing formulas of coefficients $c_1(s_1, \beta_1)$, $d_1(s_1, \beta_1)$ and the functions of curves $D_3(l_1), D_2(l_3)$ for upper surface.
3). Writing formulas of coefficients $c_2(x_p, s_2, \beta_2), d_2(x_p, s_2, \beta_2)$ and the functions of curves $D_1(l_0), D_3(l_2), D_2(l_4)$ for lower surface.
4). In this part the unknowns are found. To solve the system of five equations is used the function $Find(x_p, s_1, s_2, \beta_1, \beta_2)$ of Mathcad.
5). Writing the main functions of the upper and lower surface.
6). The formulas to calculate the circle C_s.
7). The draft of wing section is shown in Fig.1.
8). The graphs of main functions $U_1(x), U_2(x)$ and $K_1(x), K_2(x)$ are shown in Fig.2 and Fig.3, in which the graph of $K_2(x)$ has a horizontal area which corresponds the $D_1(l_0)$ curve in the nose of lower surface.
9). A fragment of the wing section is shown in Fig.4, in which the circle C_s is built and points $0, P, S_1, S_2$ are shown.
10). Printed the table of points coordinates of upper and lower surface.

1. Parameters: $r := 0.01$ $xM := 0.35$ $yM := 0.09$ $xm := 0.375$ $ym := -0.015$

2. Upper Surface $\varepsilon := 10^{-4}$ $\mu1(s1) := xM^3 - (xM - s1)^3$

$$d1(s1, \beta1) := -\frac{1 - \dfrac{xM}{r} - \left(1 - \dfrac{1}{r} - \sin(\beta1)\right) \cdot xM^2}{\mu1(s1) - \left[1 - (1 - s1)^3\right] \cdot xM^2}$$

$$c1(s1, \beta1) := -\frac{1}{xM^2} \cdot \left(1 - \frac{xM}{r} + d1(s1, \beta1) \cdot \mu1(s1)\right)$$

D3(L1)

$u1(x, s1, \beta1) := 1 - \dfrac{x}{r} + c1(s1, \beta1) \cdot x^2 + d1(s1, \beta1) \cdot x^3$ $k1(x, s1, \beta1) := \dfrac{d}{dx} u1(x, s1, \beta1)$

$$y1(x, s1, \beta1) := \sqrt{2 \cdot r \cdot \varepsilon} \cdot \left(1 - \frac{\varepsilon}{2 \cdot r}\right) + \int_{\varepsilon}^{x} \frac{u1(x, s1, \beta1)}{\sqrt{1 - u1(x, s1, \beta1)^2}} \, dx$$

D2(L3)

$u3(x, s1, \beta1) := u1(x, s1, \beta1) - d1(s1, \beta1) \cdot (x - s1)^3$ $k3(x, s1, \beta1) := \dfrac{d}{dx} u3(x, s1, \beta1)$

$$y3(x, s1, \beta1) := y1(s1, s1, \beta1) + \int_{s1}^{x} \frac{u3(x, s1, \beta1)}{\sqrt{1 - u3(x, s1, \beta1)^2}} \, dx$$

3. Lower Surface $\lambda2(xp) := \left(\dfrac{xm - xp}{1 - xp}\right)^2$ $\mu2(xp, s2) := (xm - xp)^3 - (xm - s2)^3$

$$d2(xp, s2, \beta2) := \frac{1 - \dfrac{xm}{r} - \left(1 - \dfrac{1}{r} + \sin(\beta2)\right) \cdot \lambda2(xp)}{\mu2(xp, s2) - \left[(1 - xp)^3 - (1 - s2)^3\right] \cdot \lambda2(xp)}$$

$$c2(xp, s2, \beta2) := \frac{1}{(xm - xp)^2} \cdot \left(1 - \frac{xm}{r} - d2(xp, s2, \beta2) \cdot \mu2(xp, s2)\right)$$

D1(L0) $y0(x) := -\sqrt{r^2 - (r - x)^2}$ $u0(x) := -1 + \dfrac{x}{r}$ $k0 := \dfrac{1}{r}$

D3(L2) $u2(x, xp, s2, \beta2) := -1 + \dfrac{x}{r} + c2(xp, s2, \beta2) \cdot (x - xp)^2 + d2(xp, s2, \beta2) \cdot (x - xp)^3$

$$k2(x, xp, s2, \beta2) := \frac{d}{dx} u2(x, xp, s2, \beta2)$$

$$y2(x,xp,s2,\beta2) := y0(xp) + \int_{xp}^{x} \frac{u2(x,xp,s2,\beta2)}{\sqrt{1 - u2(x,xp,s2,\beta2)^2}} \, dx$$

D2(L4)
$$u4(x,xp,s2,\beta2) := u2(x,xp,s2,\beta2) - d2(xp,s2,\beta2)\cdot(x - s2)^3$$

$$k4(x,xp,s2,\beta2) := \frac{d}{dx} u4(x,xp,s2,\beta2)$$

$$y4(x,xp,s2,\beta2) := y2(s2,xp,s2,\beta2) + \int_{s2}^{x} \frac{u4(x,xp,s2,\beta2)}{\sqrt{1 - u4(x,xp,s2,\beta2)^2}} \, dx$$

4. Decision of Equations

$$\rho(xp,s1,s2,\beta1,\beta2) := \frac{y2(s2,xp,s2,\beta2) - y1(s1,s1,\beta1)}{\sqrt{1 - u2(s2,xp,s2,\beta2)^2} + \sqrt{1 - u1(s1,s1,\beta1)^2}}$$

$$H(xp,s1,s2,\beta1,\beta2) := s2 - s1 + (u2(s2,xp,s2,\beta2) + u1(s1,s1,\beta1))\cdot\rho(xp,s1,s2,\beta1,\beta2)$$

$$k := 4 \quad x\mu := \frac{1}{2}\cdot(xM + xm) \quad y\mu := \frac{1}{2}\cdot(yM + ym) \quad xp := r\cdot\left(1 - \cos\left(2\cdot\operatorname{atan}\left(\frac{k\cdot y\mu}{x\mu - r}\right)\right)\right)$$

$$s1 := r \quad \beta1 := 0 \quad s2 := s1 \quad \beta2 := 0$$

Given $\quad y3(xM,s1,\beta1) - yM = 0 \quad y3(1,s1,\beta1) = 0 \quad H(xp,s1,s2,\beta1,\beta2) = 0$
$$y4(xm,xp,s2,\beta2) - ym = 0 \quad y4(1,xp,s2,\beta2) = 0$$

$$\begin{bmatrix} xp \\ s1 \\ s2 \\ \beta1 \\ \beta2 \end{bmatrix} := \operatorname{Find}(xp,s1,s2,\beta1,\beta2) \quad \begin{array}{ll} xp = 0.00379 & s1 = 0.01541 \quad \beta1 = -0.094 \\ & s2 = 0.02179 \quad \beta2 = 0.051 \end{array}$$

$$ys1 := y1(s1,s1,\beta1) \quad us1 := u1(s1,s1,\beta1)$$
$$yp := y0(xp) \quad ys2 := y2(s2,xp,s2,\beta2) \quad us2 := u2(s2,xp,s2,\beta2)$$

5. Main Functions

$$Y1(x) := \begin{vmatrix} y1(x,s1,\beta1) & \text{if } 0 \le x < s1 \\ y3(x,s1,\beta1) & \text{if } s1 \le x < 1 \\ 0 & \text{if } x = 1 \end{vmatrix} \quad Y2(x) := \begin{vmatrix} y0(x) & \text{if } 0 \le x < xp \\ y2(x,xp,s2,\beta2) & \text{if } xp \le x \le s2 \\ y4(x,xp,s2,\beta2) & \text{if } s2 \le x < 1 \\ 0 & \text{if } x = 1 \end{vmatrix}$$

$$\chi1(x,s1) := \begin{vmatrix} 0 & \text{if } x < s1 \\ 1 & \text{otherwise} \end{vmatrix} \quad \chi2(x,s2) := \begin{vmatrix} 0 & \text{if } x < s2 \\ 1 & \text{otherwise} \end{vmatrix} \quad x := 0, 0.002 .. 1$$

$$U1(x) := u1(x,s1,\beta1) - d1(s1,\beta1)\cdot(x-s1)^3\cdot\chi1(x,s1)$$

$$U2(x) := \begin{vmatrix} u0(x) & \text{if } 0 \le x < xp \\ u2(x,xp,s2,\beta2) - d2(xp,s2,\beta2)\cdot(x-s2)^3\cdot\chi2(x,s2) & \text{otherwise} \end{vmatrix}$$

$$K1(x) := k1(x,s1,\beta1) - 3\cdot d1(s1,\beta1)\cdot(x-s1)^2\cdot\chi1(x,s1)$$

$$K2(x) := \begin{vmatrix} k0 & \text{if } 0 \le x < xp \\ k2(x,xp,s2,\beta2) - 3\cdot d2(xp,s2,\beta2)\cdot(x-s2)^2\cdot\chi2(x,s2) & \text{otherwise} \end{vmatrix}$$

6. Circle Cs $\rho := -\dfrac{ys2 - ys1}{\sqrt{1-us2^2} + \sqrt{1-us1^2}}$ $\xi os := s1 + \rho\cdot us1$ $\eta os := ys1 - \rho\cdot\sqrt{1-us1^2}$

$\xi s(\theta) := \xi os + \rho\cdot\cos(\theta)$ $\eta s(\theta) := \eta os + \rho\cdot\sin(\theta)$ $\theta := 0, \dfrac{\pi}{50} .. 2\cdot\pi$

7. Airfoil $r = 0.01$ $xM = 0.35$ $yM = 0.09$ $xm = 0.375$ $ym = -0.015$

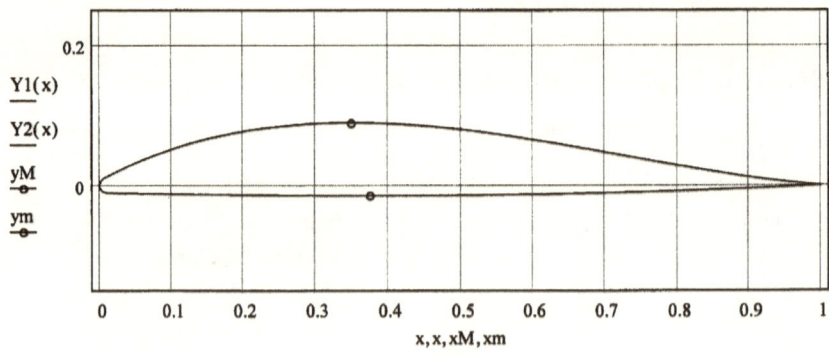

Fig. 1

8. Functions U1(x), U2(x), K1(x), K2(x) $x := 0, 0.0001 .. 0.05$

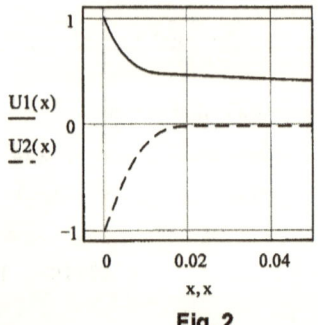

Fig. 2

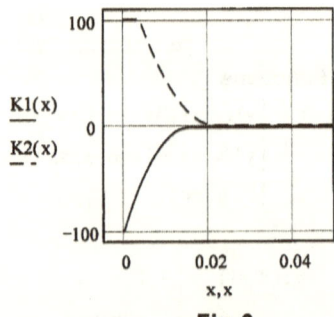

Fig. 3

9. Head of Airfoil

$x := 0, 0.0002 .. 0.07$

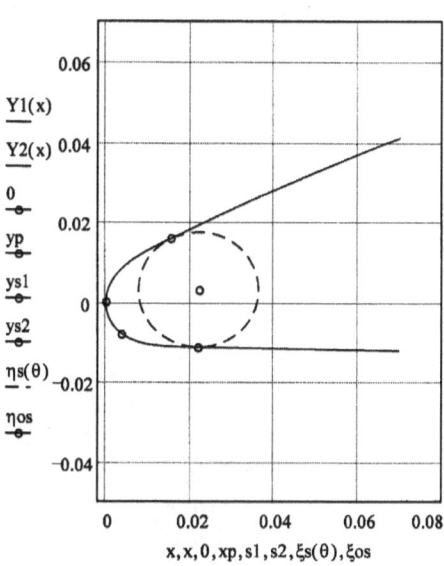

Fig. 4

10. Coordinates of Points

$$\text{Yupper}(x) := \begin{vmatrix} 0 & \text{if } |Y1(x)| < 10^{-5} \\ Y1(x) & \text{otherwise} \end{vmatrix} \qquad \text{Ylower}(x) := \begin{vmatrix} 0 & \text{if } |Y2(x)| < 10^{-5} \\ Y2(x) & \text{otherwise} \end{vmatrix}$$

$x := 0, 0.1 .. 1$

x	Yupper(x)	Ylower(x)
0	0	0
0.1	0.0521	-0.0126
0.2	0.0775	-0.014
0.3	0.0887	-0.0148
0.4	0.0888	-0.015
0.5	0.0805	-0.0145
0.6	0.0661	-0.0132
0.7	0.0482	-0.0112
0.8	0.0294	-0.0084
0.9	0.0125	-0.0046
1	0	0

x := 0, 0.002 .. 1   **Aerofoil # 1**

r = 0.01   xM = 0.35   yM = 0.08   xm = 0.3   ym = -0.02

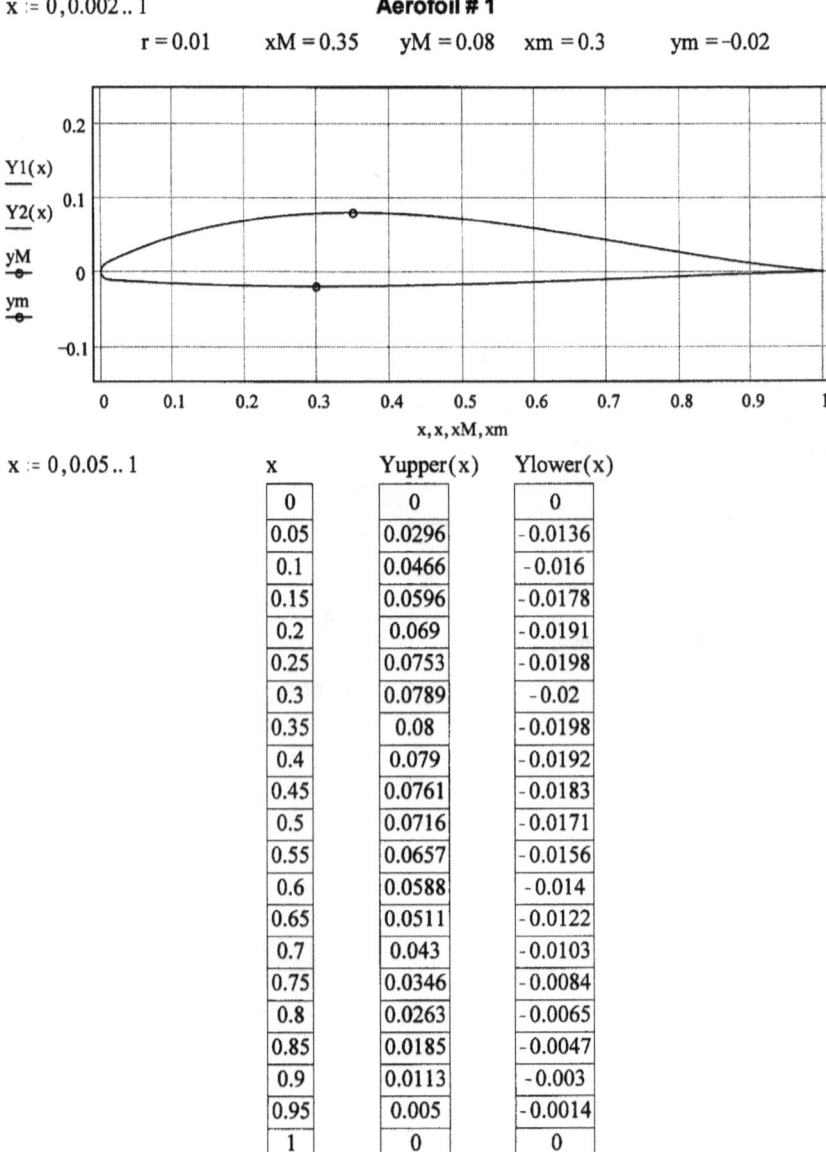

x := 0, 0.05 .. 1

x	Yupper(x)	Ylower(x)
0	0	0
0.05	0.0296	-0.0136
0.1	0.0466	-0.016
0.15	0.0596	-0.0178
0.2	0.069	-0.0191
0.25	0.0753	-0.0198
0.3	0.0789	-0.02
0.35	0.08	-0.0198
0.4	0.079	-0.0192
0.45	0.0761	-0.0183
0.5	0.0716	-0.0171
0.55	0.0657	-0.0156
0.6	0.0588	-0.014
0.65	0.0511	-0.0122
0.7	0.043	-0.0103
0.75	0.0346	-0.0084
0.8	0.0263	-0.0065
0.85	0.0185	-0.0047
0.9	0.0113	-0.003
0.95	0.005	-0.0014
1	0	0

MATHEMATICAL MODELING OF WING SECTIONS

x := 0, 0.002 .. 1   **Aerofoil # 2**

r = 0.01   xM = 0.35   yM = 0.09   xm = 0.32   ym = -0.01

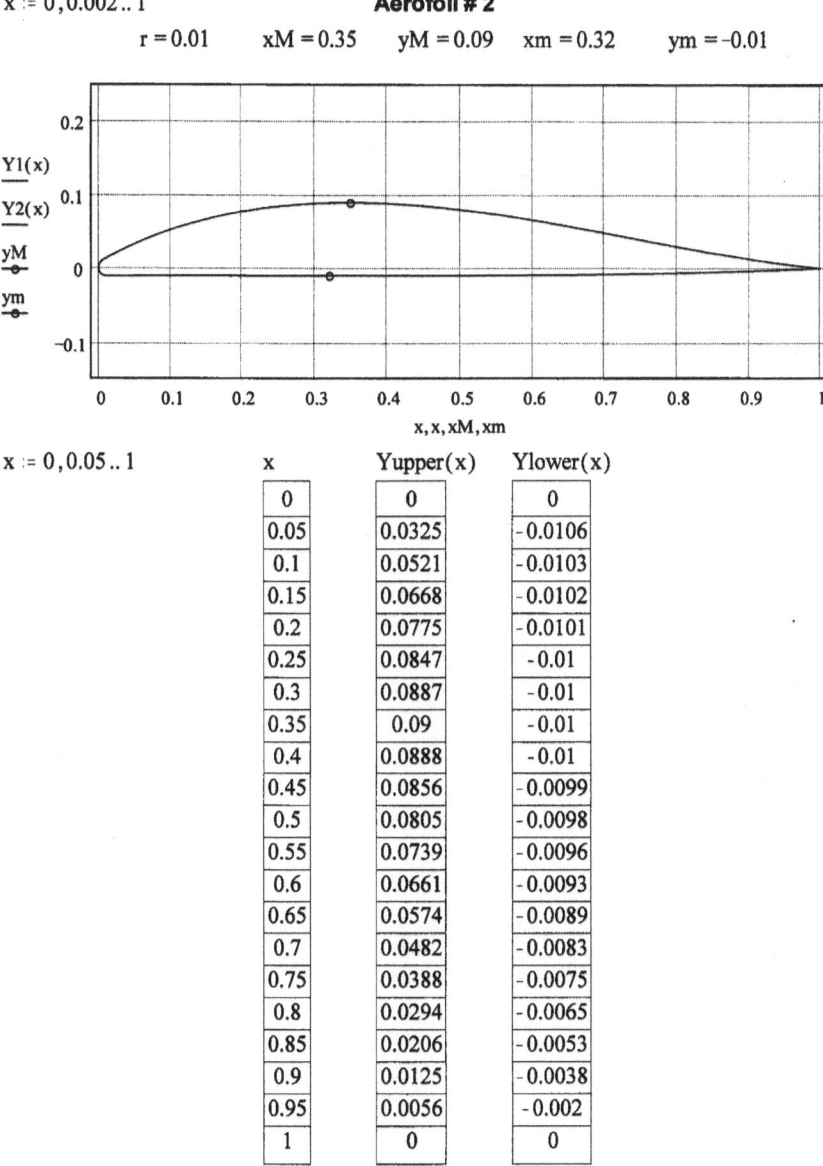

x := 0, 0.05 .. 1

x	Yupper(x)	Ylower(x)
0	0	0
0.05	0.0325	-0.0106
0.1	0.0521	-0.0103
0.15	0.0668	-0.0102
0.2	0.0775	-0.0101
0.25	0.0847	-0.01
0.3	0.0887	-0.01
0.35	0.09	-0.01
0.4	0.0888	-0.01
0.45	0.0856	-0.0099
0.5	0.0805	-0.0098
0.55	0.0739	-0.0096
0.6	0.0661	-0.0093
0.65	0.0574	-0.0089
0.7	0.0482	-0.0083
0.75	0.0388	-0.0075
0.8	0.0294	-0.0065
0.85	0.0206	-0.0053
0.9	0.0125	-0.0038
0.95	0.0056	-0.002
1	0	0

x := 0, 0.002 .. 1 **Aerofoil # 3**

r = 0.01   xM = 0.35   yM = 0.1   xm = 0.34   ym = 0

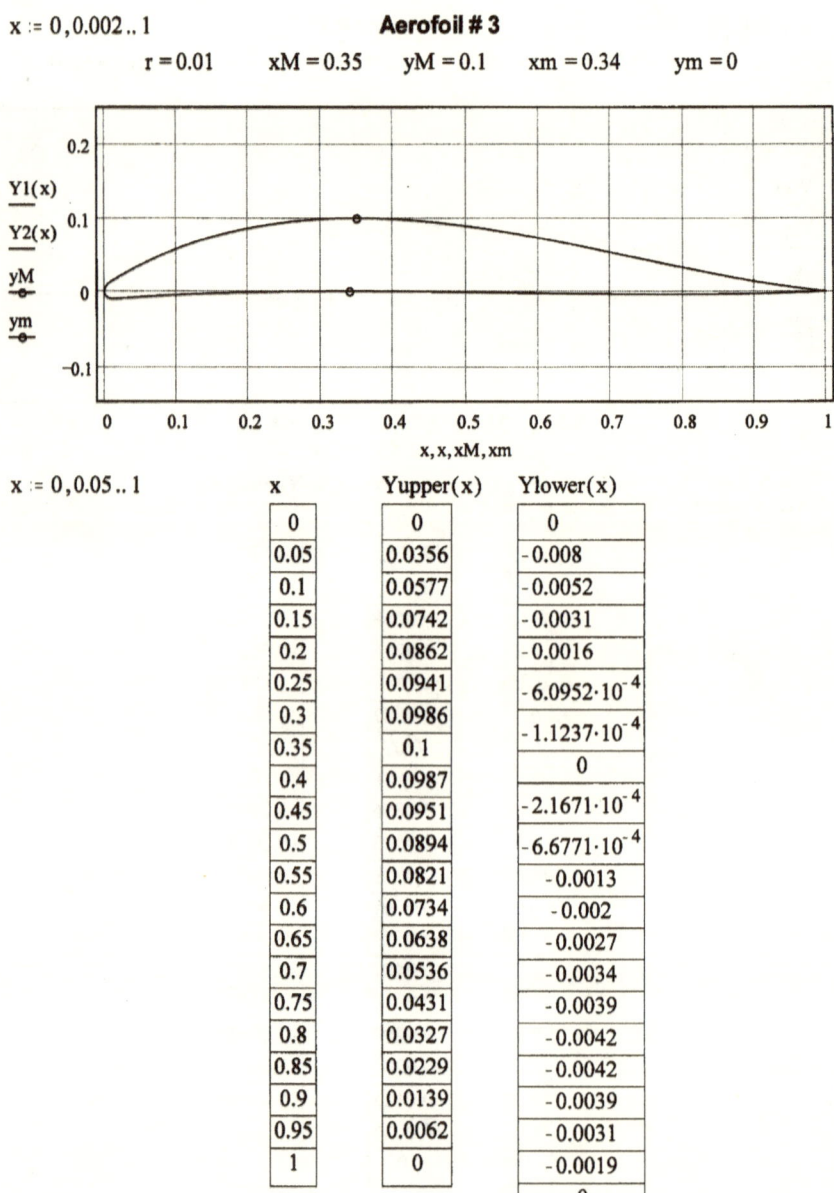

x := 0, 0.05 .. 1

x	Yupper(x)	Ylower(x)
0	0	0
0.05	0.0356	-0.008
0.1	0.0577	-0.0052
0.15	0.0742	-0.0031
0.2	0.0862	-0.0016
0.25	0.0941	$-6.0952 \cdot 10^{-4}$
0.3	0.0986	$-1.1237 \cdot 10^{-4}$
0.35	0.1	0
0.4	0.0987	$-2.1671 \cdot 10^{-4}$
0.45	0.0951	$-6.6771 \cdot 10^{-4}$
0.5	0.0894	-0.0013
0.55	0.0821	-0.002
0.6	0.0734	-0.0027
0.65	0.0638	-0.0034
0.7	0.0536	-0.0039
0.75	0.0431	-0.0042
0.8	0.0327	-0.0042
0.85	0.0229	-0.0039
0.9	0.0139	-0.0031
0.95	0.0062	-0.0019
1	0	0

x := 0, 0.002 .. 1   **Aerofoil # 4**

r = 0.01   xM = 0.35   yM = 0.11   xm = 0.36   ym = 0.01

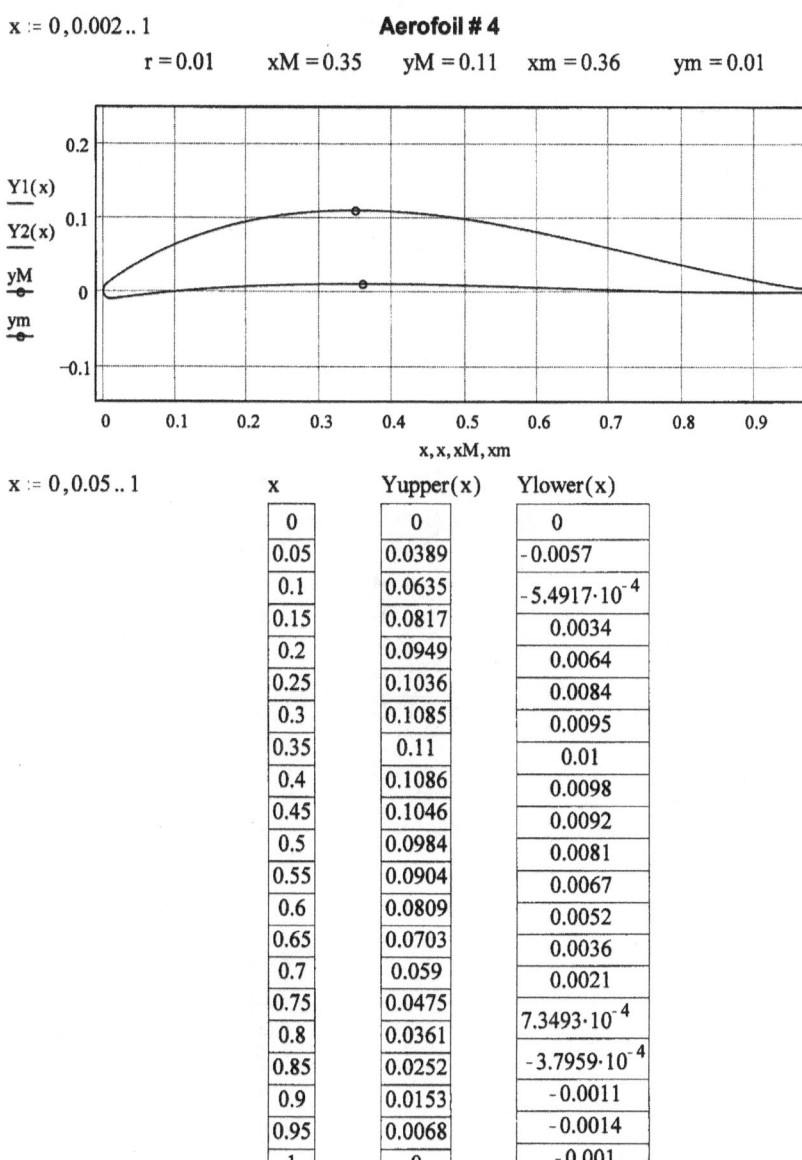

x := 0, 0.05 .. 1

x	Yupper(x)	Ylower(x)
0	0	0
0.05	0.0389	-0.0057
0.1	0.0635	$-5.4917 \cdot 10^{-4}$
0.15	0.0817	0.0034
0.2	0.0949	0.0064
0.25	0.1036	0.0084
0.3	0.1085	0.0095
0.35	0.11	0.01
0.4	0.1086	0.0098
0.45	0.1046	0.0092
0.5	0.0984	0.0081
0.55	0.0904	0.0067
0.6	0.0809	0.0052
0.65	0.0703	0.0036
0.7	0.059	0.0021
0.75	0.0475	$7.3493 \cdot 10^{-4}$
0.8	0.0361	$-3.7959 \cdot 10^{-4}$
0.85	0.0252	-0.0011
0.9	0.0153	-0.0014
0.95	0.0068	-0.001
1	0	0

$x := 0, 0.002 .. 1$

Aerofoil # 5

$r = 0.01$ $xM = 0.35$ $yM = 0.12$ $xm = 0.38$ $ym = 0.02$

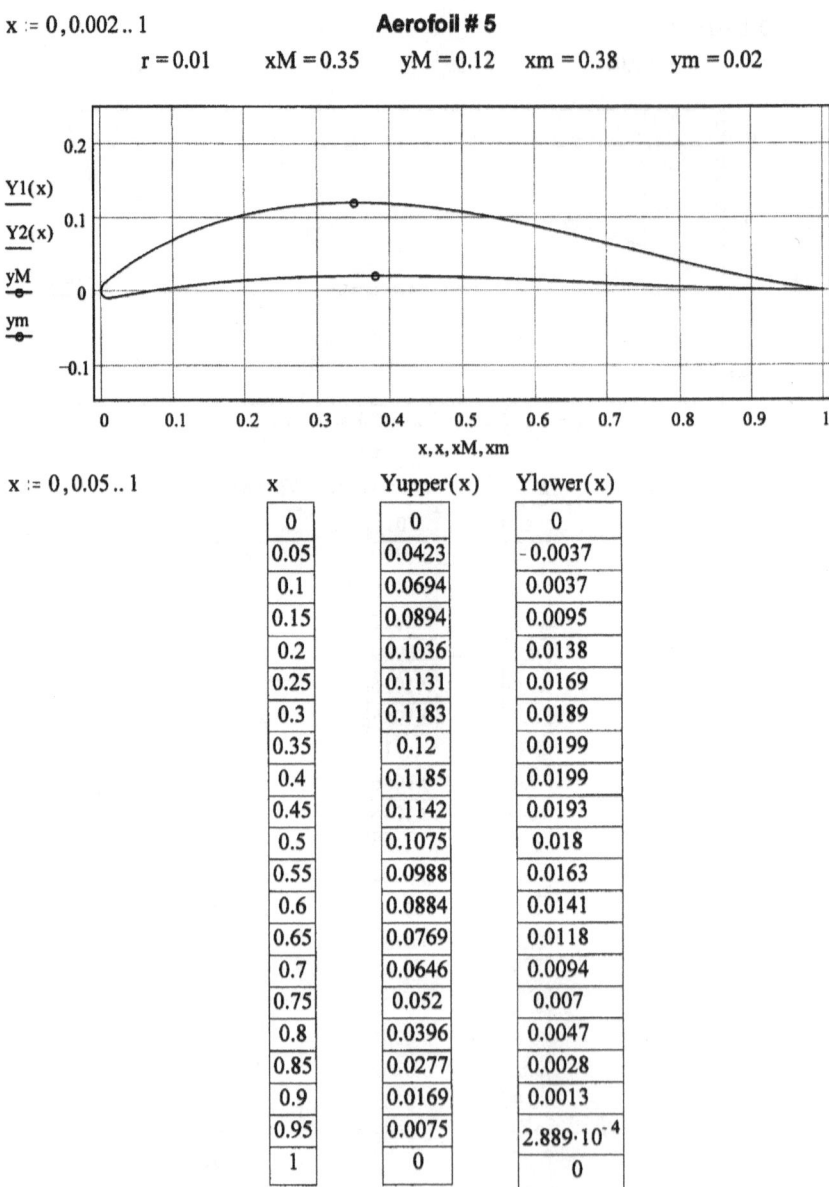

$x := 0, 0.05 .. 1$

x	Yupper(x)	Ylower(x)
0	0	0
0.05	0.0423	-0.0037
0.1	0.0694	0.0037
0.15	0.0894	0.0095
0.2	0.1036	0.0138
0.25	0.1131	0.0169
0.3	0.1183	0.0189
0.35	0.12	0.0199
0.4	0.1185	0.0199
0.45	0.1142	0.0193
0.5	0.1075	0.018
0.55	0.0988	0.0163
0.6	0.0884	0.0141
0.65	0.0769	0.0118
0.7	0.0646	0.0094
0.75	0.052	0.007
0.8	0.0396	0.0047
0.85	0.0277	0.0028
0.9	0.0169	0.0013
0.95	0.0075	$2.889 \cdot 10^{-4}$
1	0	0

x := 0, 0.002 .. 1   **Aerofoil # 6**
          r = 0.01   xM = 0.35   yM = 0.13   xm = 0.4   ym = 0.03

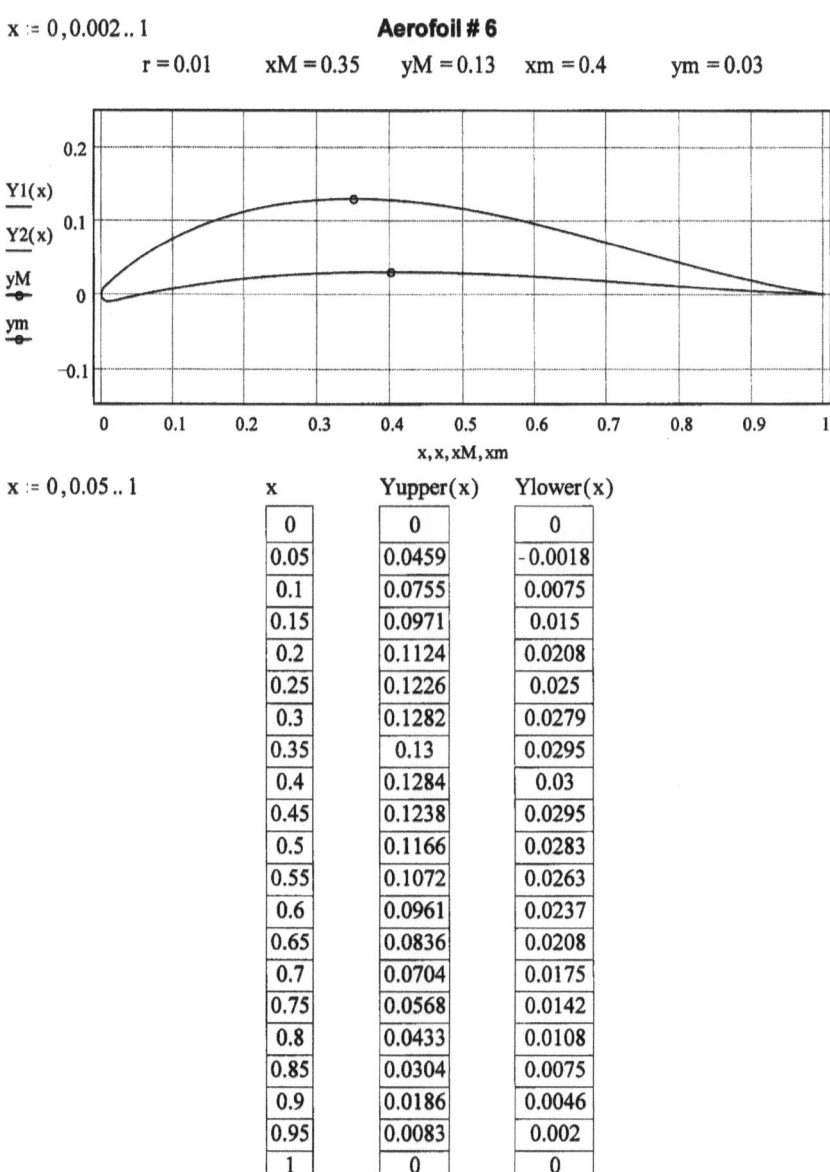

x := 0, 0.05 .. 1

x	Yupper(x)	Ylower(x)
0	0	0
0.05	0.0459	-0.0018
0.1	0.0755	0.0075
0.15	0.0971	0.015
0.2	0.1124	0.0208
0.25	0.1226	0.025
0.3	0.1282	0.0279
0.35	0.13	0.0295
0.4	0.1284	0.03
0.45	0.1238	0.0295
0.5	0.1166	0.0283
0.55	0.1072	0.0263
0.6	0.0961	0.0237
0.65	0.0836	0.0208
0.7	0.0704	0.0175
0.75	0.0568	0.0142
0.8	0.0433	0.0108
0.85	0.0304	0.0075
0.9	0.0186	0.0046
0.95	0.0083	0.002
1	0	0

www.ingramcontent.com/pod-product-compliance
Lightning Source LLC
Chambersburg PA
CBHW021049180526
45163CB00005B/2348